CUTTING-EDGE TECHNOLOGY

ARTIFICIAL INTELLIGENCE

BY BETSY RATHBURN

BELLWETHER MEDIA · MINNEAPOLIS, MN

EPIC

EPIC

Action and adventure collide in **EPIC**. Plunge into a universe of powerful beasts, hair-raising tales, and high-speed excitement. Astonishing explorations await. Can you handle it?

This edition first published in 2021 by Bellwether Media, Inc.

No part of this publication may be reproduced in whole or in part without written permission of the publisher. For information regarding permission, write to Bellwether Media, Inc., Attention: Permissions Department, 6012 Blue Circle Drive, Minnetonka, MN 55343.

Library of Congress Cataloging-in-Publication Data

LC record for Artificial Intelligence available at https://lccn.loc.gov/2019059256

Text copyright © 2021 by Bellwether Media, Inc. EPIC and associated logos are trademarks and/or registered trademarks of Bellwether Media, Inc.

Editor: Kieran Downs   Designer: Josh Brink

Printed in the United States of America, North Mankato, MN.

TABLE OF CONTENTS

LOST AND FOUND	4
WHAT IS ARTIFICIAL INTELLIGENCE?	6
HOW IT WORKS	8
HISTORY	12
TECHNOLOGY OF TOMORROW	18
GLOSSARY	22
TO LEARN MORE	23
INDEX	24

Lost and Found

A traveler stands on a busy street corner. He is lost. But his phone's map **app** will help. **Routes** appear on the screen. The traveler chooses the fastest one. Artificial intelligence helps him find his way!

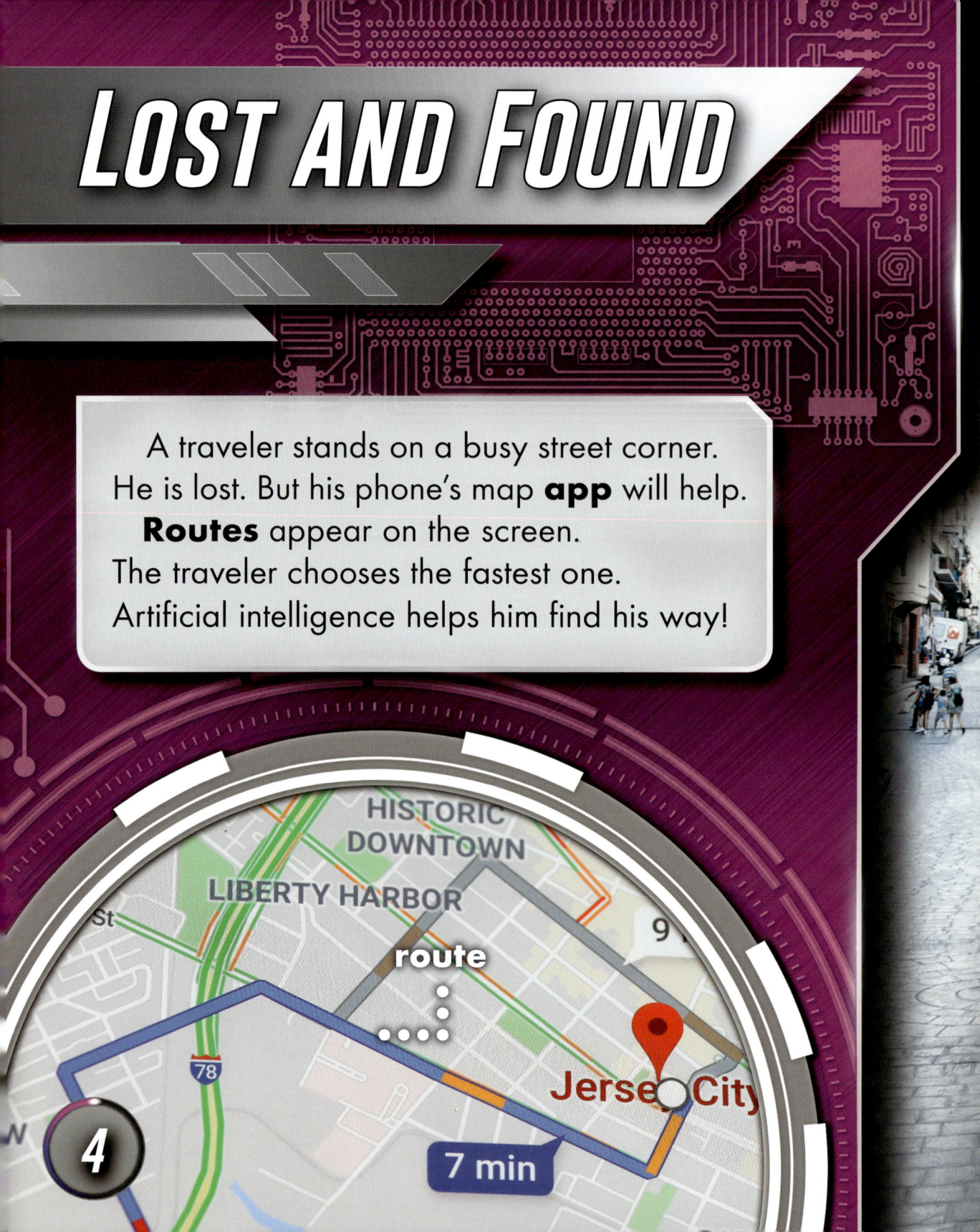

SMART PHONES

Smartphone users use AI every day. Phones do voice searches. They suggest routes. They can even recognize faces!

smartphone

What Is Artificial Intelligence?

Artificial intelligence is a computer technology. It is also called AI. It allows computers to learn and make decisions.

hospital AI

Hello How are you today?

Who Uses It?

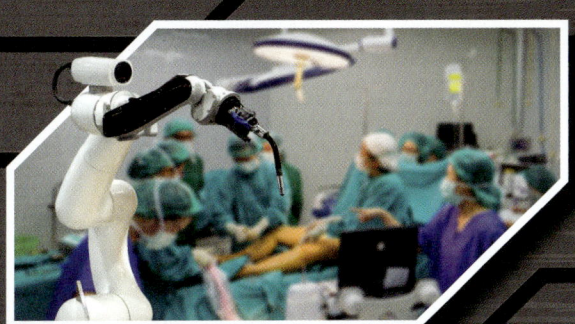
health workers

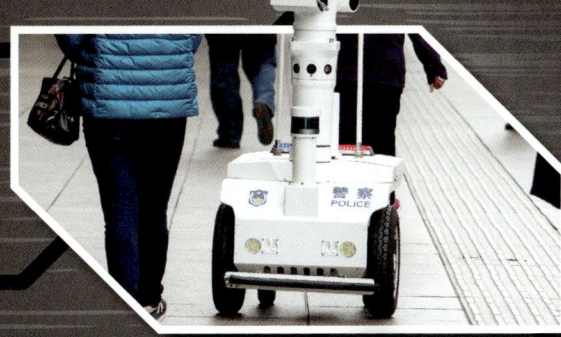
police

stores

teachers

AI is found in classrooms and hospitals. It is even found on smartphones!

How It Works

Most AI is made to do a certain job. **Programmers** give it a set of steps called an **algorithm**.

People **input** questions or requests. The AI follows the steps to do its job!

programmer

programmable robot

Some AI uses **machine learning**. Programmers teach it with training **data**. The AI creates a **model**. This helps it make guesses.

More data is added. AI uses it to make better guesses!

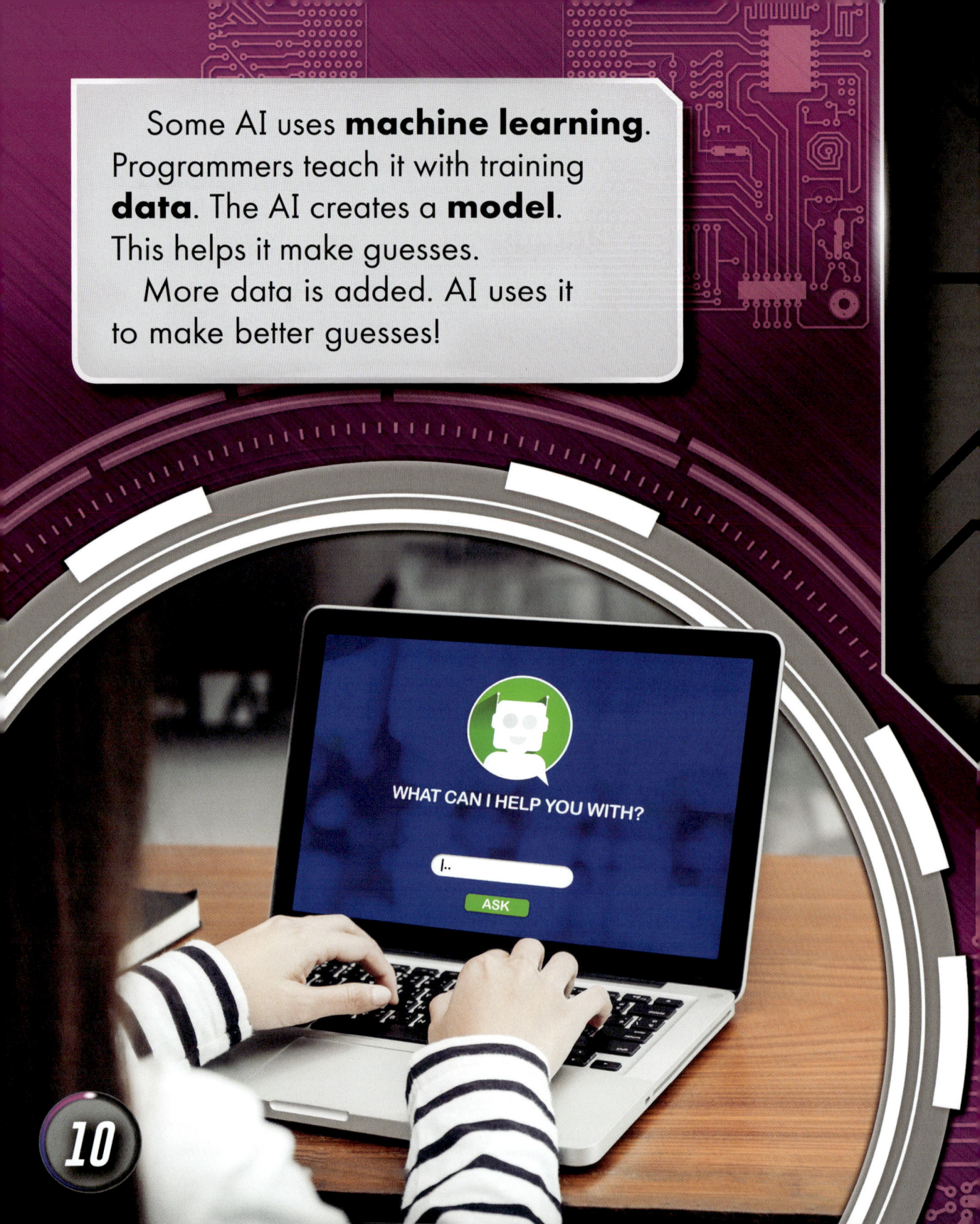

How Machine Learning Works

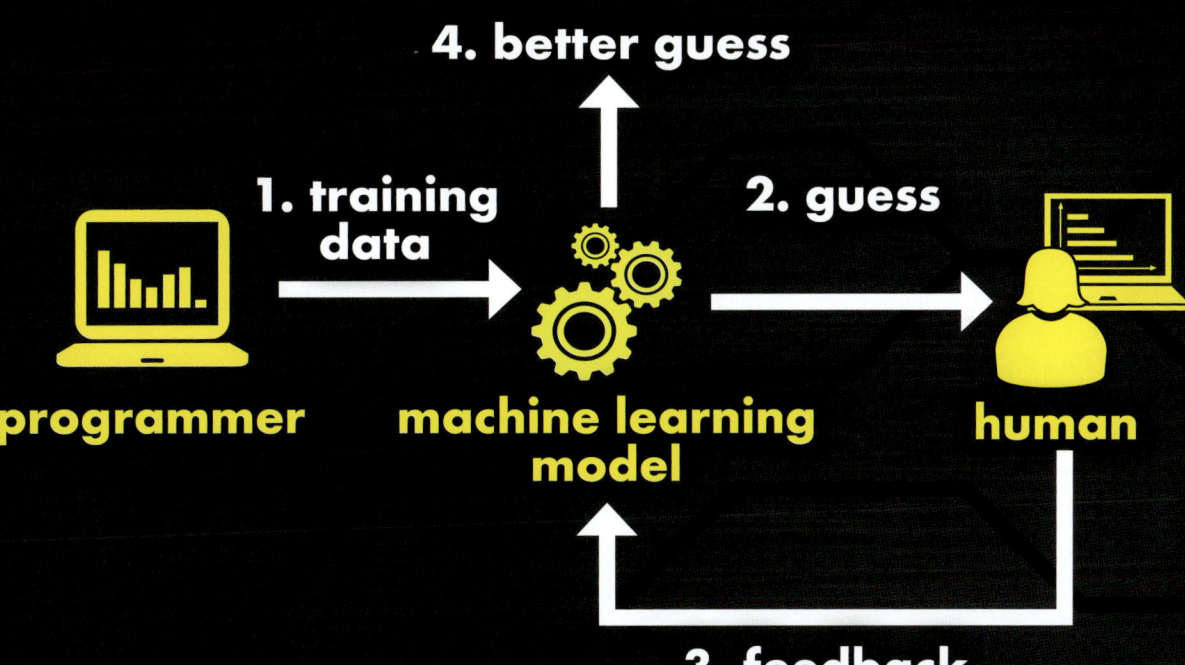

DEEP LEARNING

Deep learning is machine learning. But it uses much more data. It allows AI to teach itself!

History

Alan Turing wrote about "thinking machines" in the 1930s. This led to AI research!

Marvin Minsky researched AI in the 1950s. He thought AI should be pre-programmed. This **top-down AI** was built with all the information it needed.

Alan Turning

Marvin Minsky

TOP DOWN TALKING

Top-down AI is useful for simple tasks. For example, it might understand speech. But it does not always understand slang!

13

Artificial Intelligence Timeline

1930s — Alan Turing writes about "thinking machines"

1952 — Christopher Strachey creates the first working AI program

1950s — Marvin Minsky leads research on top-down AI

1955 — John McCarthy is the first to use the term "artificial intelligence"

Things changed in 1990. Rodney Brooks thought AI should use **neural networks**. They connect thousands of computers!

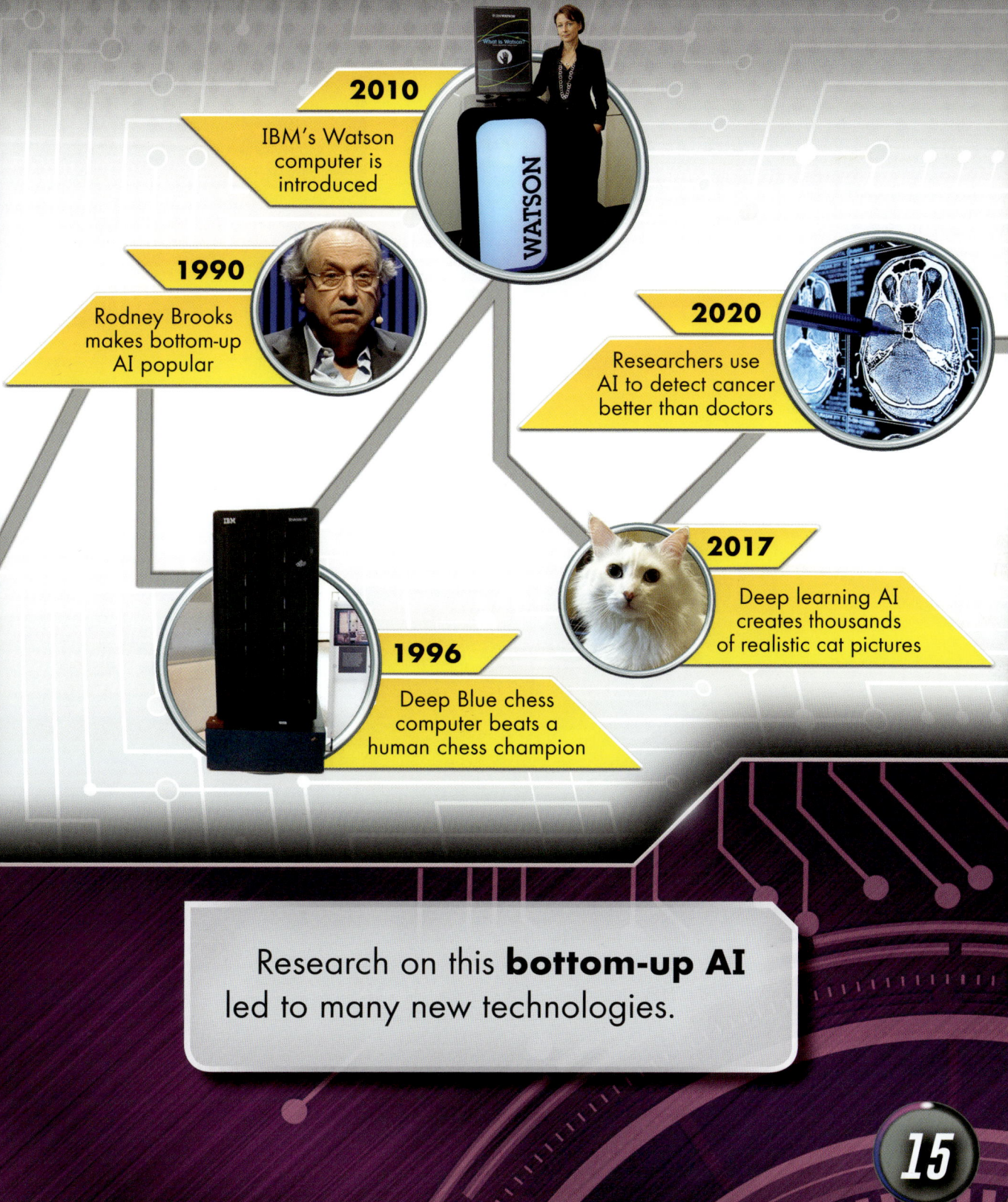

AI continued to get better. In 2010, IBM's Watson computer used bottom-up AI. It answered questions and made decisions.

Watson

AI AND JOBS

Do scientists think AI will lead to more jobs lost or more jobs created?

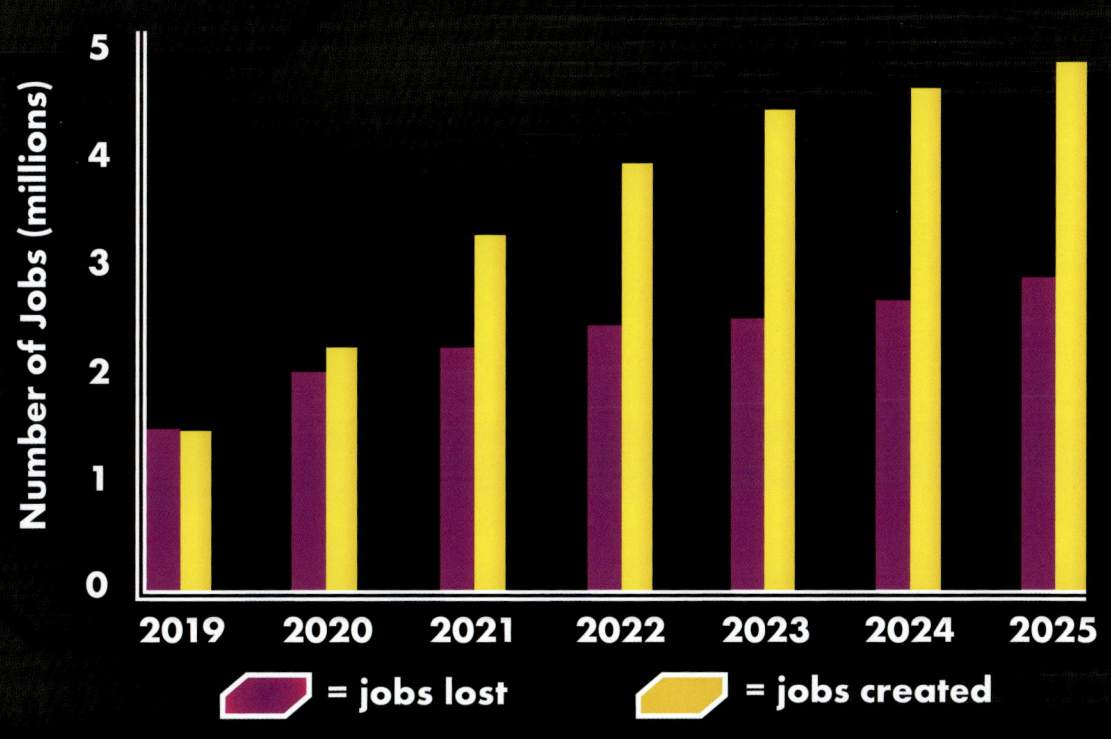

In 2012, AI got creative. The Iamus computer wrote a piece of music!

Technology of Tomorrow

More AI advances are around the corner. AI may help cars drive themselves. It may help doctors find diseases faster.

AI may help scientists explore the deep ocean. It may even help humans explore planets beyond our own!

Pros and Cons

Pros

helps humans explore

makes jobs easier

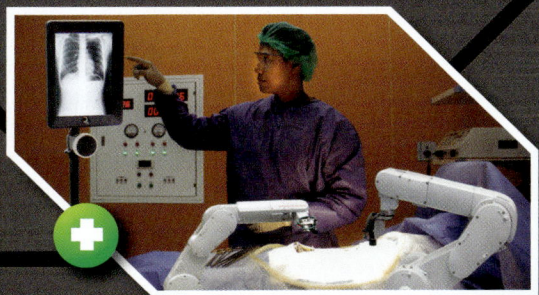

helps doctors treat patients

Cons

could become too powerful

could take away privacy

could take away jobs

Some researchers worry about AI advances. It could leave people without jobs. It could also make data less private.

Artificial intelligence is here to stay. How will it change the world?

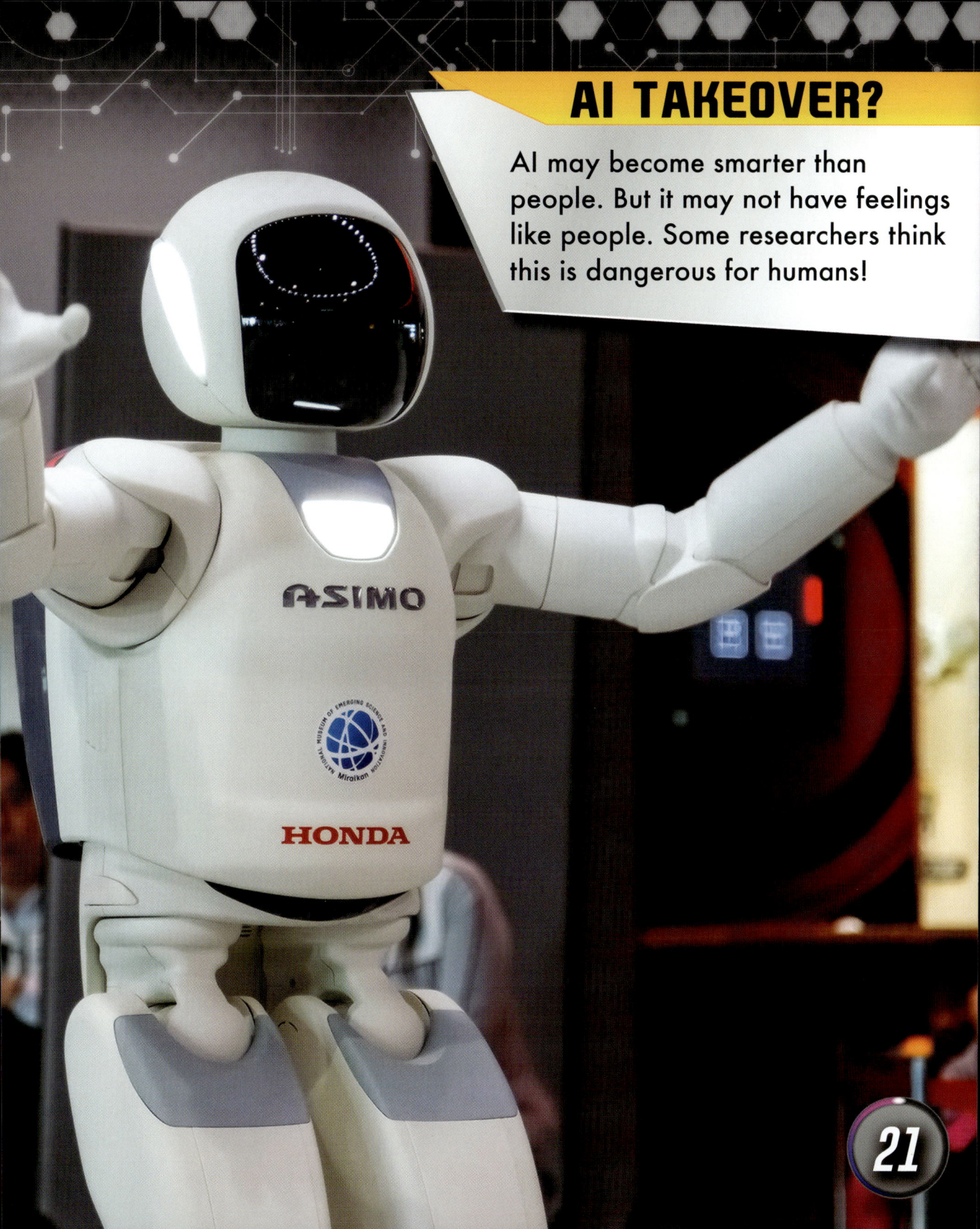

AI TAKEOVER?

AI may become smarter than people. But it may not have feelings like people. Some researchers think this is dangerous for humans!

GLOSSARY

algorithm—a set of steps that a computer follows to do its job

app—a small, specialized program downloaded onto smartphones and other mobile devices

bottom-up AI—a type of AI in which computers teach themselves from huge amounts of data

data—information

input—to put data into a computer

machine learning—a process through which computers are able teach themselves by adding new data to data they already know

model—an example used by an algorithm to make future guesses based on data learned in the past

neural networks—groups of many connected computers; neural networks are designed to work like human brains.

programmers—people who create and test programs for computers

routes—ways taken to get to other places

top-down AI—a type of AI in which computers are pre-programmed with the data that will help them do their jobs

To Learn More

AT THE LIBRARY

Dickmann, Nancy. *Robots and Artificial Intelligence*. New York, N.Y.: Gareth Stevens Publishing, 2020.

Enz, Tammy. *Artificial Intelligence and Entertainment*. North Mankato, Minn.: Capstone Press, 2019.

Gregory, Josh. *Artificial Intelligence*. Ann Arbor, Mich.: Cherry Lake Publishing, 2018.

ON THE WEB

Factsurfer.com gives you a safe, fun way to find more information.

1. Go to www.factsurfer.com.

2. Enter "artificial intelligence" into the search box and click 🔍.

3. Select your book cover to see a list of related content.

INDEX

advances, 18, 20
algorithm, 8
app, 4
bottom-up AI, 15, 16
Brooks, Rodney, 14
cars, 18
classrooms, 7
computer, 6, 14, 16, 17
data, 10, 11, 20
deep learning, 11
doctors, 18
guesses, 10
history, 12, 14, 15
hospitals, 6, 7
Iamus, 17
IBM, 16
input, 8
jobs, 8, 17, 20

learn, 6
machine learning, 10, 11
Minsky, Marvin, 12, 13
model, 10
neural networks, 14
programmers, 8, 10
pros and cons, 19
research, 12, 15, 20, 21
routes, 4, 5
scientists, 18
smartphone, 4, 5, 7
timeline, 14-15
top-down AI, 12, 13
training, 10
Turing, Alan, 12
users, 5, 7
Watson, 16

The images in this book are reproduced through the courtesy of: Tinxi, cover; Anton Gvozdikov, CIP; TY Lim, p. 4; Grusho Anna, p. 5; MONOPOLY919, p. 6; kung_tom, p. 7 (top left); atiger, p. 7 (top right); David Cardinez, p. 7 (bottom left); Monkey Business Images, p. 7 (bottom right); Pira25, p. 8; allOver images, p. 9; weedezign, p. 10; pianodiaphragm, p. 11; Heritage Image Parternship Ltd, p. 12; Hank Morgan / Rainbow, p. 13; World History Archive, p. 14 (top left); Dan McCoy / Rainbow, p. 14 (bottom left); Raymond Kleboe, p. 14 (top right); null0, p. 14 (bottom right); Steve Jurvetson, p. 15 (top left); James the photographer, p. 15 (bottom left); Daniel Maurer, p. 16; S. Bachstroem, pp. 18, 19 (top right); sdecoret, p. 19 (top right); onurdongel, p. 19 (middle left); metamorworks, p. 19 (middle right); PaO_STUDIO, p. 19 (bottom left); evakerrigan, p. 19 (bottom right); Stefano Mazzola / Awakening, p. 20; coward_lion, p. 21.